TOXIC WASTE

MALCOLM E. WEISS

TOXIC WASTE

CLEAN-UP OR COVER-UP?

Franklin Watts
New York | London | Toronto | Sydney | 1984
An Impact Book

Photographs courtesy of
UPI: pp. 5, 36, 41, 46, 66, 68, 76;
The Bettmann Archive, Inc.: p. 13;
Wide World Photos: p. 56.

Library of Congress Cataloging in Publication Data

Weiss, Malcolm E.
Toxic waste.

(An Impact book)
Includes index.
Summary: Examines the increasing problem of
hazardous waste disposal in our society.
1. Hazardous wastes—Environmental aspects—
United States—Juvenile literature.
2. Factory and trade waste— Environmental aspects—
United States—Juvenile literature.
[1. Hazardous wastes—Environmental aspects.
2. Factory and trade waste—Environmental aspects]
I. Title.
TD811.5.W45 1984 363.7'28 83-23391
ISBN 0-531-04755-5

Printed in the United States of America
6 5 4 3

CONTENTS

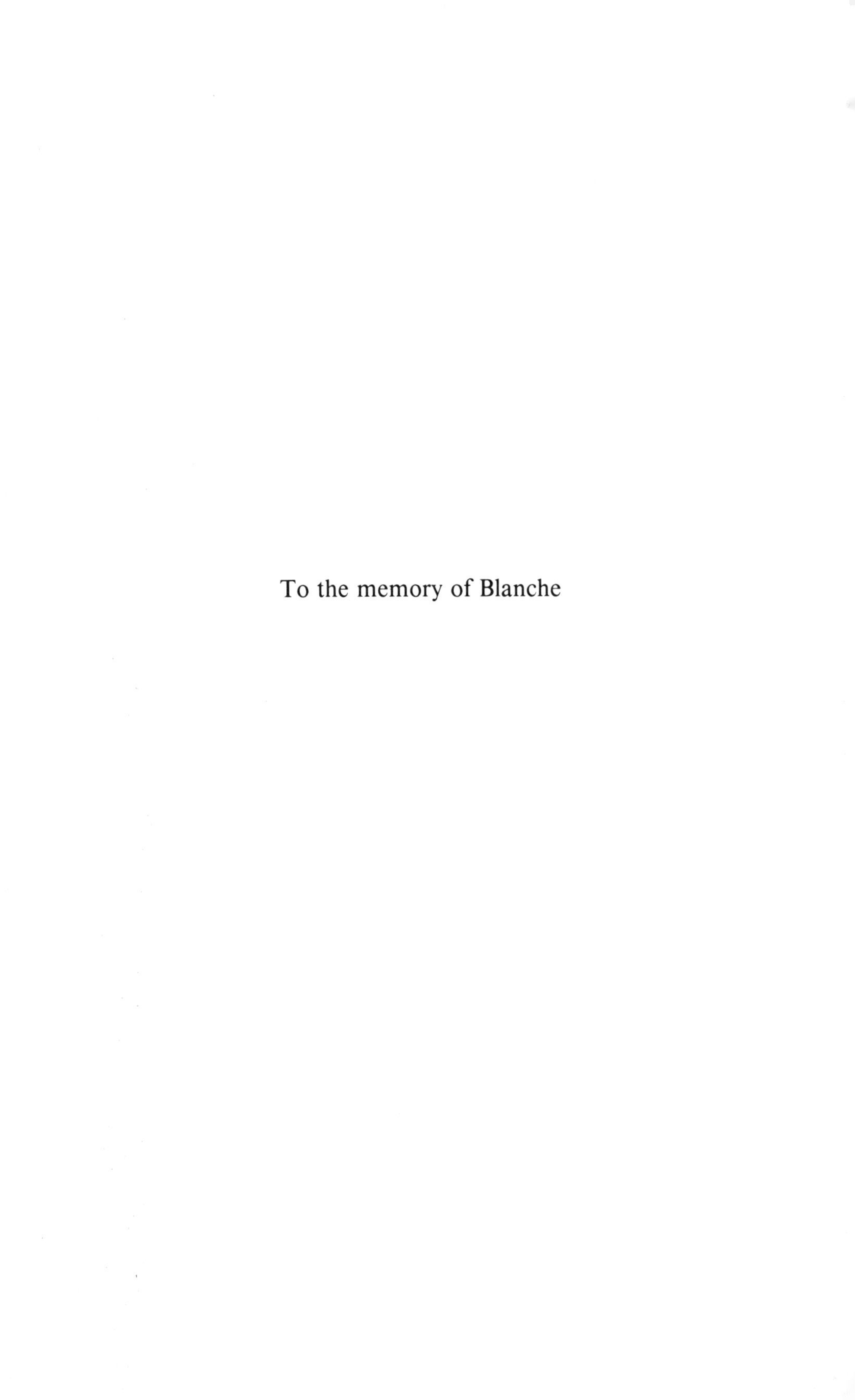

To the memory of Blanche

TOXIC WASTE

CHAPTER ONE

"TICKING TIME BOMBS"

Take about 800 tons of flour.

Mix with water to make up 400,000 gallons. Stir, if you have the strength.

Add generous helpings of cheese, tomato paste, mushrooms, and pepperoni.

Don't forget the onions!

What have you got?

You might have the makings of a pizza big enough to feed an army. Then again, you might not.

What the small town of Wellston, Ohio, had was a horrendous mess of slippery, smelly gunk—and a clogged sewage system. It was not a good start to the year 1983.

About a year earlier, a company named Jeno's, Inc., had opened a frozen pizza plant in Wellston. The plant provided 1,000 new job openings, a healthy boost for a town of 6,100 people. In return for the jobs, town officials promised that the Wellston sewage system would handle the liquid waste water from the plant.

Water is one thing. But the town officials had not planned for the huge amounts of dough, grease, and other garbage that came with the water. Over the months, pizza wastes slowly filled the sewage system's two settling tanks.

Normally, solids in sewage water settle to the bottoms of these tanks. They form a half-solid, half-liquid mass called sludge. Bacteria break down harmful substances in the sludge. Often, gas produced in this process is used as fuel to help run the sewage plant. The processed sludge can be dried and used as fertilizer. Or it may be incinerated.

But the problem with the pizza plant was that it produced too much sludge too fast. By the start of 1983, the sludge was overflowing the settling tanks. It was filling the sewage system's holding lagoon—a small pond for temporary waste storage.

The Ohio Environmental Protection Agency feared the sludge would overflow to a nearby creek and foul drinking water supplies. The agency threatened to close the plant.

Town officials were desperate. What could they do?

Could they just bury the sludge in a landfill? No, said the town's safety-service director, Richard E. Devlin. The greasy, doughy, semiliquid sludge was so slippery it might move underground, once buried. It might pollute the water supplies of nearby Columbus.

Other experts agreed with Devlin. They said the sludge must first be dried out so bacteria could act on it. Then the sludge could be buried.

That meant adding special equipment to the Wellston sewage system—at a cost of $500,000. Town officials got the money from a United States government block grant.

Wellston had had a big problem. The town would have been in real trouble if it had not gotten the money from the grant.

But by the end of 1983, Wellston's pollution problem was under control. Of course, there were headlines such as "The Pizza Is Coming!" and jokes about the pizza that ate Wellston. The idea of pizza garbage clogging a sewage system is a natural for jokes. But the Wellston story also gave people a few minutes of relief from really bad news—bad news about hazardous wastes far more dangerous and far harder to get rid of than pizza sludge.

Just before the year began, the United States Environmental Protection Agency (EPA) came out with a list of the 418 most dangerous hazardous waste dumps in the country. These dumps are scattered across the country from Maine to Florida, from the East Coast to the West—and beyond. Some are located on islands in the Pacific Ocean that are owned or held in trust by the United States government. Only four states did not appear on the list: Alaska, Hawaii, Nevada, and Wisconsin.

But the EPA list was far from complete. Experts estimate there are somewhere between 12,000 and 15,000 such dumps throughout the United States. The EPA's most-dangerous list only includes dumps that are no longer being used. It does not include dumps still receiving wastes. Furthermore, it does not include *any* hazardous waste dumps on property owned by the federal government, such as land at military bases.

When the EPA released its list of 418 dumps, the head of the agency called the sites "ticking time bombs." Even as she spoke, one of those "time bombs" was exploding. Compared to an ordinary bomb, it was a slow explosion. In fact, it's still going on. It will be years before we know how much damage has been done.

In Missouri, the winter of 1982–3 was wet and warm. There were floods. On December 5, 1982, the Meramec River, a small branch of the Mississippi, flooded out the town of Times Beach. About 2,200 people were forced to leave their homes. Weeks later, when the flood waters receded, only two hundred came back.

Why? Samples of soil from Times Beach showed that it was contaminated with a highly toxic substance—dioxin.

Dioxin was found in the soil in parts of Times Beach about a week before the flood. After the flood, there was fear that contaminated soil carried by the floodwaters might have spread over the whole town.

Back came the investigators from the EPA. They were dressed in silvery-white air-and-waterproof protective suits. They breathed through special masks that filtered all hazardous substances out of the air. Around them, children and grownups who had come back to Times Beach walked and played, wearing ordinary clothing.

"It's real crazy," said one EPA investigator. "We walk into these people's homes with our suits on, and their kids are just running around." He added that the EPA crew must look like the agents for the fictional United States space agency who stalked the Extraterrestrial through a California home in the movie *E.T.*

In and around Times Beach, the investigators found concentrations of dioxin ranging from twenty parts per billion (ppb) to one hundred ppb. Scientists at the United States Centers for Disease Control in Atlanta, Georgia, say that the maximum safe level for dioxin is one part per billion. That makes dioxin one of the most poisonous substances known.

Let's get a clearer idea of how small one ppb really is. To get one ppb of just one ounce of dioxin, you would have to dilute that ounce with 7,812,500 *gallons* of water. That is a lot of water. Suppose a family of four used water at the huge rate of 400 gallons a day—100 gallons for each family member. It would take them fifty-three years to use up 7,812,500 gallons. And if all that water

Technicians from the Environmental Protection Agency take soil samples in Times Beach, Missouri, where high levels of the toxic chemical dioxin had been found.

contained more than one ounce of dioxin, it might well damage their health.

Look at it another way. Dilute one drop of dioxin with about 13,200 gallons of water. That's 1 ppb of dioxin.

As we've seen, the town of Times Beach had dioxin levels 20 to 100 times above the "safe" level. The danger seemed so great that the EPA offered to buy the entire town from its residents. That would give people who owned homes there the money to make a new start somewhere else.

But Times Beach is only part of Missouri's hazardous waste problem. By mid-1983, there were fifteen known dioxin-contaminated areas in the state. An additional eighty sites were being investigated. On one road in St. James, about one hundred miles southwest of St. Louis, dioxin levels were found to be 1,800 ppb.

Toxic as it is, dioxin is only one of many man-made hazardous wastes. The problem is nationwide—and worldwide. The "time bombs" have been exploding for a good many years.

• In 1972, large amounts of poisonous hexachlorobenzene (HCB) were dumped into a landfill in Louisiana, in cattle-farming country. HCB evaporates easily. Cattle over a 100-square-mile area breathed in the vapor. They had to be destroyed.

• In the same year, scientists found that some 1,500 large metal drums of toxic wastes had been buried near the town of Byron, Illinois. These drums had accumulated for years. No one knew who had dumped them. They contained cyanides, poisonous metals such as lead and mercury, corrosive acids, and other dangerous substances. Wildlife and stream life were destroyed. All plants were killed over a wide area. The EPA estimates it will take many years, and hundreds of millions of dollars, to clean the site and repair the damage.

• In 1978, a fire broke out at a waste dump in Chester, Pennsylvania. Fifty thousand drums of waste had been stored there over a three-year period. Forty-five firemen were overcome by the chemical fumes which caused lung, skin, and eye damage. Homes in the area were evacuated. Later investigation showed that wastes had been dumped directly on the site without even being buried. Some were seeping into the nearby Delaware River.

• People, especially children, in the neighborhood of Love Canal in Niagara Falls, New York, became seriously ill because of exposure to chemicals, including dioxin, buried there twenty-five years earlier. Over two hundred families were evacuated from Love Canal in 1978 and 1979. About eighty different chemicals, many of them known to be cancer-causing, were found buried in the neighborhood.

• Much the same problem exists in Europe and around the world. Many European rivers are badly polluted by industrial wastes. Oil, poisonous metals, and other toxic substances have been poured into the Mediterranean Sea for decades. The wastes come from nations in Europe, the Middle East, and Africa that border on the Mediterranean.

Lev Temerman, a Soviet scientist who now lives in Israel, described a huge wasteland near the Russian city of Kyshtym. In 1960, Temerman was driving along a highway near the city. He saw ". . . a large area . . . in which people were evacuated and villages razed . . . to prevent inhabitants from returning, there was no agriculture or livestock raising, fishing and hunting were forbidden."

Scientists think there was a nuclear explosion in the area of Kyshtym. The cause of the explosion is not clear.

It may have been an explosion at a nuclear power plant, or at a plant that processes nuclear wastes for reuse. Or it may have been caused by careless disposal of nuclear wastes. Whatever happened, it is clearly another instance of the growing problem of hazardous wastes.

Why has the problem of toxic wastes become so great and so widespread? How did it get started?

CHAPTER TWO

THE GOOD OLD DAYS?

In a sense, the problem of hazardous wastes began with the beginning of life. Living things produce wastes. They must get rid of them. If wastes pile up in an organism, they kill it. This is true of all living things, from single-celled creatures to higher plants and animals.

But what is waste for one kind of living thing is useful to another. Carbon dioxide is a waste product for animals. It is vital to plants. Plants use water and carbon dioxide and solar energy to make starches and sugars. In the process, plants release oxygen into the air. Animals breathe in oxygen. In their body cells, oxygen is combined with starches and sugars. This is a kind of slow, controlled fire without flame. It makes energy to help run the body. The waste products of this fire—the "ashes"—are carbon dioxide and water. It is those ashes that plants use to make the basic foods for most other living things, as well as oxygen. And so the cycle goes on and on.

This is only one of the many cycles of life. The remains and wastes of both animals and plants are used by bacteria and fungi, some of the earth's simplest living things. Bacteria and fungi use some of the digested wastes as food. They return others to the soil. These wastes in turn can now be used by plants.

So living things depend on each other in very complex ways. It is not just a matter of eating and being eaten. Wastes produced by some living things are put to use by other living things. Sooner or later, all wastes are recycled and used over and over again.

This has been true since life began. It was still true when people first appeared on the scene, millions of years ago. It has remained true throughout most of human history—until the last century or so.

Nevertheless, even in prehistoric times, people had special problems with wastes. Probably one of the earliest problems was due to the discovery of fire.

We know that many prehistoric people lived in caves. They used fires in the caves for cooking, heat, and sometimes to drive wild animals away.

What was it like to live in such caves with smoky fires? How did it affect the health of the cave dwellers?

We don't know, but we can guess. The lungs of an Egyptian mummy, thousands of years old, were found to be black with soot. And in Roman times, people often lived in deep caves in the winter with a season's supply of food. The air in those caves was almost always smoky from fires.

Like modern-day smokers, cave people may have coughed and wheezed their way through life. They may have liked it that way. After all, where there's smoke, there's fire. Fire meant warmth and food and safety. Fire meant home.

What's more, smoke may have helped to mask some of the other odors that hung in the air. They were the odors of the garbage dump: rotting food, old bones, well-gnawed and not-so-well-gnawed, human and animal wastes.

Every cave had its kitchen midden. The phrase comes from a Danish word that means "kitchen dung heap."

But kitchen middens were much more than mere garbage heaps. Broken or worn-out tools and weapons were thrown on the pile. So were bits of plants and flowers used for medicines and food.

The middens built up into layers. Today, scientists can read the history of the cave-dwellers in those layers.

Evidently, the cave-dwellers lived with their kitchen middens and liked it. The piled-up waste may have caused problems with disease, and perhaps even led to occasional epidemics, but people survived.

At first, the problems could not have been too bad. Only small numbers of people lived in each cave. People moved from place to place as they hunted. Much of their time was spent in the open air.

Then prehistoric people discovered farming. They began to raise cattle, sheep, and other animals. They planted crops.

Farming provided a larger and more dependable supply of food. That meant more people could live together in one place. People found they could stay in that place and tend their crops year after year, century after century. Where there was a good supply of water and rich soil, more people lived together and there were more farms. Villages and towns grew up. People invented tools for better farming.

Farmers also need to keep track of time and the seasons so that they know when to plant and when to harvest. People watched the movements of sun, stars, moon, and planets and learned to make calendars. Special days on the calendar, like the start of spring, became religious holidays. Slowly, laws were written down, and governments formed. Civilization began.

People had made a lot of progress since the days when they lived in caves. But in one way, things had not changed at all. People still dumped their garbage wherever they wanted to. And that was usually in the easiest and nearest place.

Most people lived in simple one-room huts with dirt floors and no furniture. If they were farmers, they shared the room with poultry and pigs. The fact that people and animals were crowded together in villages made the problem of wastes worse than it had been in the caves.

At best, some villages had a dump at the edge of the village where garbage could be left. Even so, people preferred to dump most garbage on land near their homes. That was where they farmed. Through thousands of years of experience, farmers had found out that most wastes somehow made soil better for growing things. On the other hand, people did not see any connection between epidemics of disease and the filthiness of their villages. Little was done to clean them up because no one saw a need for it.

The filth got worse as cities developed, several thousand years B.C. The population of early cities was in the hundreds or thousands. Later, that grew to millions. But numbers aside, the filth and unpleasantness of a city like eighteenth-century London differed little from the conditions of cities thousands of years earlier. The streets were unpaved. They were narrow lanes of dirt with a gutter in the middle. People tossed garbage and the contents of their chamber pots out the windows onto the streets. Sometimes they warned passers-by that a shower of waste was coming. Sometimes they did not.

Frequent rains helped to some extent to clean the London streets. The filth ran through the gutters in streams and poured into the River Thames. Nobody worried

This engraving by the English artist William Hogarth, shows a street scene in eighteenth-century London—when waste was carelessly poured out the window and refuse choked the streets.

about that. At London, the Thames was too salty to drink from anyway.

The scene would have been much the same in a city in ancient Greece. The main differences: There would have been less filth because there were fewer people. And instead of being muddy filth it would have been dusty filth, because the climate of Greece is drier than that of London.

There were a few remarkable exceptions among ancient cities. Rome, for example, had paved streets and a well-constructed water supply and sewage system. It also had a system of public baths. Ancient Romans bathed more frequently than any later generation right up to the end of the nineteenth century.

Until that time, most people regarded the Roman use of water as just another example of their love of luxury. Here and there, a doctor spoke out about contaminated water supplies as a major source of epidemic disease. But these warnings were generally ignored. Life went on. The people who lived through one epidemic did not get quite so sick the next time a similar epidemic came along. However filthy the cities got, the filth was eventually broken down and returned to the soil. For the cities' wastes were the same kind of natural wastes that had been in the world since prehistoric—and prehuman—times.

But that was to change.

CHAPTER THREE

HOW WE GOT TO WHERE WE ARE

The ancient Romans were almost modern in their use of paved streets and water systems. It seems likely that they were "modern" in another way, too. Many Romans may have been the victims of slow poisoning. They seem to have been poisoned by a hazardous metal—lead.

Lead does not exist as a metal in nature. The most common ore of lead is the mineral galena. Galena is lead sulfide. There are rich underground deposits of galena and other lead ores in the United States, Canada, Australia, and the Soviet Union. The ore is dug up and heated with charcoal and lime to produce metallic lead.

Lead was mined and produced in much the same way in ancient Rome. The word "galena" is Latin for "lead ore."

Lead is a soft metal that can readily be molded into different shapes. It is also very resistant to rusting in air or water. When it is exposed to either, a thin film of lead rust (lead oxide) forms on its surface. This film protects the lead beneath from further rusting.

For this reason, the Romans used lead for water pipes. Many of their pipes are usable today, after nearly 2,000 years. The Latin inscriptions on the pipes are still easy to read.

Lead is a poison that the body can get rid of only very slowly. Suppose someone gets a small amount of lead with his or her food or water each day. Over the months and years, the poison builds up in that person's body. Slowly, symptoms of lead poisoning develop.

The symptoms involve the nerves. Lead injures and eventually kills nerve cells. As the lead accumulates in the body, it also interferes with the normal working of the brain. The victim's behavior becomes odd. He or she may have sudden violent fits of temper or of maniacal excitement. There may be convulsions. Parts of the body may be numbed or paralyzed.

Some scientists point out that several Roman emperors, among them Claudius, Caligula, and Nero—showed symptoms like these.

Some of this poisoning may have come from the lead in Roman water pipes, but that was probably not its main source. The Romans drank a lot of wine. The wine was fermented in lead pots. Romans also sweetened their wine with syrup. The syrup was made by simmering grape juice—again in lead pots. The same kind of syrup was used as a sweetener in other Roman foods. A modern-day Canadian scientist, Dr. Jerome O. Nriagu, has remarked, "One teaspoon of such syrup would have been more than enough to cause chronic lead poisoning."

The idea that the Roman upper class may have suffered from lead poisoning is not a new one. Nor is it accepted by all scientists and historians. However, Dr. Nriagu found another clue that has given weight to the theory. He discovered that many of these Romans also had gout—a painful disease of the joints.

One kind of gout is caused by lead poisoning. According to Dr. Nriagu, Roman records show that this is the kind of gout that was common in Imperial Rome.

Mercury is another metal that has been known since

ancient times. It has been found in Egyptian tombs that date back to 1500 B.C.

Mercury is the only metal that is a liquid at ordinary temperatures. Ancient peoples thought of it as liquid silver, or quicksilver. "Quick," in this sense, means "living." Mercury was thought to be a kind of silver so pure that it flowed like a liquid and seemed alive.

Like lead, mercury is produced from its ores. The main ore is cinnabar. Cinnabar itself makes a brilliant red waterproof pigment called vermillion. Vermillion is one of the oldest pigments used by people for cosmetics and dyes. Mercury itself was valued in ancient times as a magical substance.

Like lead, mercury is a nerve poison that builds up slowly in the body. The symptoms of mercury poisoning are similar to those of lead poisoning.

Mercury is a more powerful poison than lead. The fact that it is a liquid also makes it more dangerous. It can be absorbed through the skin if it is handled. When the liquid is spilled, it breaks up into many tiny droplets. These droplets evaporate, becoming mercury vapor—mercury in the form of a gas. Even very small amounts of mercury vapor breathed in day after day will soon cause mercury poisoning.

People have known about this danger for a long time. Centuries ago, mercury miners at Almadén, Spain, were only allowed to work in the mines for thirty-two hours a month.

In other times and places, however, people have ignored the threat mercury poses. Until the 1940s, mercury poisoning was common among people who worked in hat-making factories. Up to that time, mercury compounds were used to make the felt for felt hats. Many workers got mercury poisoning. They trembled and were easily upset or angered. Eventually, some suffered severe brain damage.

People did not realize that the hatters' odd behavior

was due to mercury poisoning. Instead, it was popularly believed that hatters were often a little strange to begin with. Lewis Carroll may have had this idea in mind when he created the character of the Mad Hatter in *Alice in Wonderland.*

Mercury is no longer used for hat making. Safer substitutes have been found. Substitutes have been found for mercury used for other purposes, too. Mercury compounds were once used in the treatment of certain diseases, including syphilis. Now, safer medicines have taken their place. Some mercury-based fungicides are now banned in the United States.

The use of lead has also been banned or limited. Lead is no longer used in indoor paints. In the 1970s, some cities, including New York, passed laws that limited the amount of lead in gasoline sold there. The lead is used to prevent "knocking" of the car engine. Knocking, which is due to uneven burning of the fuel in the engine, can damage engine parts. On the other hand, cars burning leaded gas spew out lead particles and vapor from their exhausts. On crowded city streets, that means dangerous amounts of lead in the air. This is especially true during traffic jams. Cars may stay in one spot for half an hour or more with their engines running and exhaust fumes building up.

In 1965, the United States Congress amended the Clean Air Act, to limit exhaust emissions from automobiles. As a result, car makers had to add special equipment that removed harmful substances from engine exhausts before they could get into the air. It turned out that small amounts of lead stopped this equipment from working properly, so more and more engines were built which could run on unleaded gas without knocking.

However, it seemed as if no sooner had one source of lead or mercury poisoning been eliminated than others sprang up. New uses are being found for these metals faster than old uses are being discarded.

Why? Over the past century or so, scientific and medical knowledge has increased at an explosive rate. We have gone from horse-drawn carriages to railroads, automobiles, jet planes and spacecraft. Television and satellite hook-ups link all the world together. On the farm, as everywhere else, machine power has largely replaced muscle power. One reason that farms can produce more food today than ever before is that machines do most of the work. And new vaccines and medicines allow people to be healthier and to live longer than in the past. In the United States and other western countries, most people lead far better lives today than their ancestors did a century ago.

But there is a darker side to the story. To keep our highly-mechanized world running, we must rely on materials like lead and mercury. For example, lead is used in storage batteries and in nuclear reactors. Mercury has many vital electrical and chemical uses.

In addition, improved health and increased life expectancy have led to rapid population growth. There are four times as many people in the United States today as there were a hundred years ago. About 75 percent of the population live in cities and their suburbs. A century ago, 75 percent lived in rural areas.

That poses a problem for the United States and for other industrialized nations: how to dispose safely of increasing amounts of waste?

Our large cities, mechanized farms, and huge power plants all produce wastes. Many of these wastes are hazardous. As population in industrialized countries increases, production goes up—and so does the amount of waste. But the land and the water in which the waste can be disposed of does not increase. In fact, the world's supply of land and water *decreases.*

As cities grow in size, the amount of land available for dumping shrinks. You can't dump highly poisonous wastes in the middle of a city street or in the subway system. Nor can you dump them in the sewage system. They would eventually come back into the city's water supply to poison the people living there.

Throughout the twentieth century, the problem of hazardous waste disposal has increased steadily. At first, people were not aware of it—any more than the Romans, 2,000 years ago, gave thought to lead as a source of poisoning. But gradually, people realized that the problem was a serious threat to life and health. One of the first places it became noticeable was in Minimata, a city on the Japanese island of Kyushu.

In 1956, a strange disease broke out in Minimata. Its victims trembled uncontrollably and had convulsions. They were easily upset, shouting loudly. Many became deaf, dumb, and blind. Forty-six people died. Before long, the illness had been dubbed "Minimata Disease."

All the sick people were among the families of fishermen. All were in the habit of eating lots of shellfish from Minimata Bay. Were the shellfish responsible for Minimata Disease?

At first, scientists believed the answer was "Yes." They thought the shellfish must have been contaminated, perhaps with what is known as "red tide."

Shellfish eat one-celled organisms in the sea. Sometimes they feed where there are swarms of such organisms that produce powerful poisons. In eating the bacteria, the shellfish take these poisons into their own bodies.

There, the poisons are concentrated. The ability to concentrate materials is important for shellfish. Sea water is like a thin diluted soup of many chemicals that are useful to life. There are, for instance, small amounts of the metal calcium in sea water. Shellfish can extract calcium from sea water and use it to form the lime of their shells. The shell of a mussel or a clam has thousands of times more calcium in it than an equal weight of sea water does. In the same way, if sea water contains the red tide organism, shellfish living in that water will concentrate the poison.

Oddly enough, such a concentration does not harm the shellfish—but it will harm any human who eats that shellfish. Within half an hour of eating a creature contaminated with red tide, a person will become ill. The symptoms of red tide poisoning include vomiting, stomach cramps, a prickling, itching sensation around the mouth, and weakness in the muscles.

As we've seen, though, the symptoms of Minimata Disease were quite different. And those symptoms developed slowly, over a period of days or weeks. For months, their cause remained a mystery.

Then, gradually, doctors and scientists began to recognize that the symptoms of Minimata Disease resembled those of mercury poisoning. The odd behavior, the trembling, were common to both conditions. The fact that some victims of Minimata Disease slowly lost their sight and hearing seemed to confirm the idea that the disease was attacking their nervous systems. Yet the disease was far more severe than any recorded outbreak of mercury poisoning.

The questions about Minimata Disease piled up. Was it really a form of mercury poisoning? If so, what made it so severe? Where was the mercury coming from? Was there a link between Minimata Disease and the shellfish in Minimata Bay? Could there be mercury in the shellfish?

There was. And it was a very unusual form of mercury—methylmercury. Methylmercury is a mercury compound made by living things such as the bacteria that live on the bottom of Minimata Bay. These bacteria do not normally make methylmercury. But when there is mercury in the water, they absorb it with their food supply, producing methylmercury.

Tests showed that methylmercury, once eaten by people, attacks the brain directly. That explains why victims of methylmercury poisoning show such severe symptoms.

Normal brain cells resist absorbing most poisons from the blood. The body's defense against such absorption is called the "blood-brain barrier." Only after the body has been weakened by mercury poisoning over a period of time does that barrier begin to break down. Then brain damage begins. Unlike mercury, however, methylmercury passes right through the blood-brain barrier. Brain damage starts soon after the poison gets into the bloodstream.

Now the Minimata mystery was starting to unwind. The bacteria in Minimata Bay made methylmercury. The shellfish absorbed it. People ate the shellfish. Result: Minimata Disease. But where was the mercury coming from? Finding the answer was important. The disease was spreading to northern Japan.

It took scientists ten years to locate the source of the mercury in Minimata Bay. Eventually, they found that a company that manufactured plastics on the bay was dumping its mercury wastes into the water. At first, the company denied the dumping. Then, as the evidence grew, the company changed its story.

Yes, company officials said, we are dumping mercury in the bay. But we're not dumping very much. The bay is so big compared to the amount of mercury dumped. It would be like putting a tiny bit of salt into a bathtub full of water. The salt would be spread throughout the water so thinly that it would hardly be noticed.

There were a lot of things wrong with that argument. First, the company had not dumped a small amount of mercury, but a very large amount—about 400 tons. Second, the mercury was not spread evenly throughout the waters of the bay. Mercury is heavy, and most of it had settled to the bottom. Third, as we have seen, bacteria converted the poisonous substance into the even-more-dangerous methylmercury.

Executives at the plastics company were correct in saying that the methylmercury was spread out widely over the floor of the bay. However, they did not realize that the shellfish could absorb the poison and concentrate it in their bodies.

At last, the pieces of the Minimata puzzle were fitting into place. One piece was ignorance—ignorance of the way shellfish concentrate poisons, ignorance of the way a heavy metal behaves in water. Other pieces had to do with life in an industrialized society. Japan is a heavily populated nation with little room to spare for hazardous waste dumping. Yet many people in Japan are used to, and demand, the products and conveniences of modern living—the plastics, the electrical energy, the chemicals.

Industries grew up in Japan to meet worldwide needs for those products. The more efficiently and cheaply each company made its products, the greater the demand for them—and the more money the company could make.

The cheapest way to get rid of industrial waste was to dump it in the nearest stream, river, or bay. If there was no body of water nearby, the stuff could be dumped or buried on land. Few people knew or cared about the dangers of such a policy. After all, hadn't things been done that way for thousands of years?

Yes, they had. But now there were too many toxic wastes. There were too many dumps, and too many people too close to them. Sooner or later there would be real trouble. In short, Japan was facing the same dangers from hazardous wastes that the rest of the industrialized world was—and for the same reasons.

CHAPTER FOUR

WHAT ARE HAZARDOUS WASTES?

By the end of 1982, the federal Environmental Protection Agency classified some 400 chemicals as hazardous wastes. Many of these substances are *organic* compounds.

The name organic is misleading. Originally, chemists used the label *organic chemistry* to mean the chemistry of substances made by living things. These substances all contain the element carbon.

Carbon atoms can link up with each other, forming chains that may contain hundreds of thousands of carbon atoms. Atoms of other elements—especially hydrogen, oxygen, and nitrogen—combine with various carbon atoms in the chain. This creates the complex pattern of twisting, looping, folding, cross-linked chains that make up one molecule of an organic compound. The pattern is unique for each compound.

The bodies of living things are made up of many millions of complex carbon compounds. At one time, chemists thought that such compounds could be made only by living *organ*isms. Thus the compounds were called *organic.*

That distinguishes such compounds from those that do not contain carbon. These inorganic compounds are made up of a small number of atoms. (A few simple carbon compounds, such as carbon dioxide and sodium bicarbonate, are also classified as inorganic. Such carbon compounds do not contain carbon chains.)

The distinction between organic and inorganic compounds is a useful one. But as early as 1828, scientists learned how to make organic compounds in the laboratory that were identical with compounds made by living things. Before the end of the nineteenth century, chemists were going much further. They succeeded in making organic compounds that had never existed in nature—synthetic organic compounds. Today, chemical manufacturers turn out many millions of synthetic organic compounds. Some of them are made from inorganic chemicals. Many are derived from petroleum, which itself is a complex mixture of organic compounds.

These man-made organic compounds include plastics of all sorts, synthetic fibers, insecticides, herbicides, a large variety of medicines and drugs, paints, and petroleum by-products.

Petrochemical products, like other synthetic organic compounds, are essentially new "arrangements" or linkages of the molecules in petroleum. Sometimes nonorganic chemicals, such as chlorine, are also added. But all the molecules in these raw materials are not used in the finished product, and the various combinations of these leftover molecules are called by-products or waste products.

All manufacturing produces wastes. Some of these wastes are hazardous. But the chemical industries alone produce about 60 percent of all manufacturing wastes in the United States. And practically all the wastes produced in the manufacture of herbicides, insecticides, and petroleum refining are hazardous.

Dioxin is probably the most notorious hazardous waste. Actually, "dioxin" is just a shorthand name for a family of 75 synthetic organic compounds. All have similar structures. All are toxic.

In general, dioxins are unwanted by-products of a group of organic compounds called chlorinated phenols. A typical phenol is carbolic acid, a sharp-smelling corrosive substance used as a disinfectant and in the manufacture of plastics and dyes. The plastics and dyes are produced by adding chlorine atoms to the carbolic acid. Herbicides, pesticides, wood preservatives, and some pharmaceuticals are made by chlorinating other phenols.

One of the dioxins is the most toxic man-made chemical known, by the standards of many tests with different groups of animals. Its chemical name is usually abbreviated to the initials TCDD. This is the dioxin found at Times Beach, Missouri.

In the first chapter, we saw that the Centers for Disease Control told Missouri officials that the maximum safe level for TCDD in soil was one part per billion. Dr. Renate D. Kimbrough, a scientist on the CDC staff, says that this does not mean that lower concentrations of TCDD are "safe." She points out that long exposure to lower concentrations of the dioxin could still be damaging to health. Laboratory tests on animals bear this out. A daily dose of five parts per trillion (ppt) of TCDD causes a significant increase in cancer among rats.

What does "significant" mean in such tests? For a large number of test animals, such as rats, scientists know that a certain percentage will be likely to develop cancers. Suppose five ppt of TCDD is added to the diets of half these rats. Over their lifetimes, higher percentages of these rats develop cancer than do those whose diet contains no TCDD. Some increase in the cancer rate might be due to chance. No two groups of animals can be exactly alike. But an increase in the cancer rate above a certain amount is highly unlikely to be due to chance. Scientists use special statistical formulas to calculate what that amount will be for a given test.

In the test where the rats got five ppt of TCDD daily, the increase in the cancer rate was well above the chance level. This is a significant result. That means that the increase was not due to chance. It was due to some change in the rats' environment. And only one thing had changed under these carefully controlled laboratory conditions: The rats with the higher cancer rate were getting daily doses of TCDD. Therefore, the change was due to the TCDD.

Other laboratory tests show that exposure to TCDD in extremely small quantities can cause damage to the liver and nervous system. Scientists at the federal EPA write in a booklet called *Dioxins* that, "The slightest trace of TCDD in the environment may have adverse effects on the health of both human and animal populations."

Dioxin-contaminated chemicals have been manufactured in substantial amounts since the 1950s. Only in the last decade, however, have scientists become aware of the hazardous effects of the dioxins and of TCDD in particular.

How much TCDD has been produced over the past thirty years? It's impossible to get an exact figure. But Dr. Alvin Young, of the Veterans Administration, has made an estimate based on the figures released by the chemical industry. Dr. Young is a specialist in health problems among American veterans who were exposed to the dioxin-contaminated herbicide, Agent Orange, which was used extensively in the Vietnam War.

Young says the figures show that some 800,000 gallons of TCDD-contaminated products were manufactured in the United States in 1975—a typical year. From what scientists know about the proportion of dioxin contamination in these products, Dr. Young calculates that about 2.1 kilograms (4.6 pounds) of dioxin were contained in the finished products, and about 80 kilograms (176 pounds) were left over as a waste—a total of 82.1 kilograms (180.6 pounds).

Young believes that at least that amount of dioxin has been produced each year from the 1950s to about 1975. After that time, companies manufacturing dioxin-contaminated chemicals improved their methods of manufacture so the percentage of contaminating dioxin was reduced. For example, in the 1960s, the herbicide 2,4,5,-T contained as much as 70,000 ppb of TCDD. By 1983, manufacturers claimed to have reduced that to 10 ppb.

Still, if Young's figures are near the mark, about 2,053 kilograms of TCDD were produced in the United States between 1950 and 1975. That is 4,517 pounds, or more than 2¼ tons—an enormous amount, considering how toxic the substance is in microscopically small quantities. And production goes on, even if in reduced amounts.

Experts agree that large amounts of TCDD are buried in some of the thousands of hazardous waste dumps scattered around the United States. And that's about where agreement ends.

• There is disagreement among scientists on the health effects of dioxin where people have been exposed to it in relatively large amounts—in Times Beach and at Love Canal in New York, for example. Dr. Henry Falk, chief of the special studies branch at the CDC, says that TCDD contamination is "very much regarded as a significant health problem that needs to be handled with extreme care." He says that the acute (immediate) effects of TCDD on people are only part of the problem. There are long-term effects. We've seen that TCDD causes cancer in laboratory animals, for example. Similar effects on people may take twenty years or longer to show up. "There is a lot of uncertainty about the long-term effects on people," says Dr. Falk.

• Dr. Philip H. Abelson, editor of the magazine *Science,* is less uncertain about some long-term effects. He discussed the hazards of TCDD in an editorial in the June 24, 1983, issue of *Science.* He refers to an accident at a chemical plant in 1949, where 121 workers were exposed to TCDD. That was long enough ago for cancer to show up. "While ironclad proof of a null effect is missing," Abelson writes, "so too is a basis for believing that TCDD is a dangerous carcinogen in humans." In other words, there was no absolute proof that TCDD did not have a cancer-causing effect in these accident victims. But there was no proof, either, that TCDD is dangerous to people as a cancer-causing agent.

• Scientists at the National Institute of Occupational Safety and Health (NIOSH) said something quite different in a letter to the medical journal *Lancet*, dated January, 1981. They studied the cases of workers at chemical plants in Michigan and West Virginia who were exposed to TCDD. Of the 105 exposed workers who died, three developed a very rare form of cancer. Only one person out of 1,400 is expected, on average, to get this type of cancer. Three people out of 105 is a very high rate. It suggests that TCDD played a part in causing the cancers.

Actually, there is a problem with both Dr. Abelson's sample of 121 workers and the NIOSH sample of 105. Both samples are small. As a rule, the smaller the sample, the less reliable the results. We need to wait for larger samples to get more accurate measures of the long-term damage done to people by TCDD. NIOSH scientists are now studying the causes of death in 3,000 workers at plants making 2,4,5-T. That will be completed in 1985. Other studies will be going on for years longer.

And in the meantime, what do we do? Nothing?

Hardly. Gary Stein was CDC's field coordinator at

Times Beach. He was aware of all the uncertainties. He said, "We're in a gray zone, because the data aren't available to give precise risk estimates."

The animal studies pointed to the potential dangers for people. Stein and his co-workers advised the people at Times Beach to leave.

Yet as long as animal studies are the major source of information about TCDD, there will be no way to estimate exactly how risky exposure is for people. There are difficulties with using animal experiments to predict what will happen to humans. For example, it takes less than one millionth of a gram (about 4/100,000,000ths of an ounce) of TCDD to kill a 400-gram (fourteen-ounce) guinea pig. Yet in proportion, it takes 5,000 times as much TCDD to kill a hamster.

What does this tell us about how susceptible to TCDD *people* are? Not much. Scientists who say we must wait for human data before deciding what to do about hazardous wastes are quick to point to such flaws.

To other scientists, such arguments are a quibble or, worse, beside the point. Dr. Samuel S. Epstein, an expert on hazardous wastes at the University of Illinois Medical Center, says that animal experiments are the traditional means of determining human risks. If a large number of different animals show a wide variety of bodily damage from TCDD—and they do—that is a danger signal for people as well.

Direct evidence of how hazardous wastes affect people cannot come from experiments. It must come from exposures due to accident, or from exposures that took place before the substances were known to be hazardous. Such evidence, too, is often open to doubt. Suppose, for instance, a large group of people exposed to TCDD for many years get cancer. The causes of cancer are complex. What these people ate or drank over the years may have helped cause the cancer. So may many other things in their environment were possible factors—exposure to X-rays, sunlight, other chemical pollutants, and even psychological stress.

Once again, the arguments can go on and on. Some will raise serious questions. Others will be merely a means of avoiding responsibility for costly clean-ups and the need to compensate victims for the damage to their health. For similar reasons, it took scientists forty years to prove the connection between smoking and lung cancer. Today, cigarettes bear the warning that they are "dangerous to your health." Yet some scientists working for the tobacco industry still dispute the finding that smoking can cause cancer.

These same problems of determining health risks crop up with other hazardous wastes as well. Among the organic hazardous wastes are PCBs, or polychlorinated biphenyls. These compounds are related to the chlorinated phenols used as herbicides and pesticides.

PCBs have been used since the 1930s as coolants in electrical transformers. They are liquids, resistant to heat, light, and water. About one and a half billion pounds of PCBs have been manufactured in the United States since the 1930s.

Unfortunately, PCBs are highly toxic. There is evidence that they cause cancer, birth defects, and damage to skin, eyes, and lungs. PCBs leaking from junked transformers have created major hazardous waste problems. Under certain conditions, fires in PCB-filled transformers can also generate dioxins. In April, 1979, the EPA banned the manufacture of PCBs, and they are no longer made in western Europe, either. But existing PCB transformers and other products continued in use. Their safe disposal is regulated by the EPA, but PCBs disposed of in the past are still a big part of the hazardous waste problem.

Still another source of hazardous wastes are organic

solvents. A solvent is a substance used to dissolve other substances. Solvents are usually liquids. Water is the most common and frequently used solvent. But most organic compounds are not soluble in water. They are soluble in a variety of organic solvents. A familiar example is acetone, the basic ingredient in nail polish remover. Nail polish contains certain organic compounds. Naturally, they are not soluble in water—otherwise the polish would "run" on a rainy day. But nail polish is soluble in acetone. Other examples of organic solvents are carbon disulfide, benzene, and phenol. Phenol is used in refining fuel oil. The phenol removes other petroleum products, such as asphalt and wax, from the oil. Such solvents are used in manufacturing to remove impurities from the finished product.

In the past, such solvents have been discarded after use. While the solvents themselves are toxic, the impurities they contain are usually even more so. A further danger is that in a dump such solvents may dissolve other wastes and carry them into water supplies. For example, TCDD is known to stick tightly to soil particles. And TCDD is insoluble in water. Industry chemists say this means TCDD will move only slowly through the soil and will not enter the water supply. Critics reply that organic solvents in the dump can dissolve TCDD and carry it into the water supply.

Other hazardous wastes include corrosive materials such as strong acids and alkalis, explosive and flammable substances such as the by-products of oil refining, and heavy metals such as lead, mercury, and cadmium.

One kind of hazardous waste is unique. This is radioactive waste. It includes the highly radioactive waste produced at nuclear power plants and the low-level radioactive waste from hospitals. Hospitals use low-level radioactive materials in the detection and treatment of cancer and other diseases.

Radioactive wastes are unique because in a sense they are indestructible. The radioactivity which makes such wastes hazardous is a result of the unstable structure of the atoms of the radioactive substance. Many of these substances will take thousands upon thousands of years to lose enough of their radioactivity to become harmless. And there is no way to speed up the process. No physical or chemical change will alter radioactivity.

The only way to deal with such wastes is to store them. But how can materials that will be deadly for thousands of years be stored so that they will never do anyone any harm in the future?

Some scientists suggest burying them 2,000 to 3,000 feet deep at certain sites. These sites are so-called stable geologic formations. The scientists believe such formations will not shift or change for 10,000 years. There will be no release of radioactivity during that time, they say.

As an added precaution, the wastes may be solidified inside of certain ceramic materials. The wastes would actually become a part of the molecular structure of these materials. The ceramics are resistant to water and to corrosive substances, and they are very tough. This ceramic "package" itself would be placed in a stainless steel canister. The canister would be blanketed in thick layers of absorbent materials. The whole thing would be buried underground at a geologically stable site.

Other scientists are skeptical. There is no way to predict the stability of a rock formation over a period of 10,000 years, they say. And if the rocks in the burial site do shift, no packaging will withstand the resulting stress and strain. Moreover, some of the radioactive substances will be deadly for much, much longer than 10,000 years. According to one group of experts, the Union of Concerned Scientists, there is no such thing as a safe way to store some highly radioactive substances.

However, there are a number of ways that other hazardous wastes can be stored or treated. How effective are they?

The cheapest and most common method is the so-called secure landfill. It has two things in common with the burial method for radioactive wastes. In the first place, it does not dispose of hazardous chemical wastes by neutralizing them, or by destroying their poisonous properties. It just stores them away. Secondly, experts agree that the best-made landfills in existence will eventually leak. Their poisonous contents will escape. Landfills are designed to be monitored for decades so that leaks may be detected. But some toxic substances will remain poisonous for centuries. Toxic metals are permanently toxic.

The Chemical Manufacturers Association claims that most hazardous wastes can be placed in landfills that are "environmentally sound." But even they warn that we "need to reduce dependency on landfills," and suggest that the United States is "headed toward a condition where landfills will be used as a last resort."

Another way of dealing with the problem of hazardous wastes is to change manufacturing processes so that fewer wastes are produced. This is bound to be expensive. But in the long run, companies may save money by wasting fewer of their raw materials. In addition, new laws and regulations place greater responsibility on industry for treating the wastes it produces. Such laws may make industrial wastes more and more expensive to produce. Eventually, industries may find it cheaper to alter their processes.

Laws can also encourage recycling and reuse of wastes. New laws require that hazardous wastes be treated and made harmless before being disposed of. Reuse is relatively easy in the case of solvents. The waste dissolved in the solvent is removed by distillation. The solvent is used again.

Many hazardous wastes can be rendered harmless by being heated to very high temperatures. The temperatures break down the toxic molecules into harmless or even useful components. If the substance is heated in air, the process is called incineration.

Incineration can be used with all organic wastes that do not contain heavy metals. Many such wastes are broken down into carbon dioxide and water. Chlorinated organics have special problems, however. They give off vapors of hydrochloric acid, sulfur dioxide, and dioxins, threatening people who live near an incineration site.

But incineration can be done far out at sea, away from populated areas. In 1982, an incineration ship called *Vulcanus* was allowed by the EPA to incinerate 3.6 million gallons of PCBs. *Vulcanus* was built by Waste Management Inc., of Oak Brook, Illinois. It is fitted with two incinerators and can burn 4,000 gallons of waste an hour. Reportedly, *Vulcanus* destroyed over 99 percent of the PCBs it burned in 1982. The hydrochloric acid it produced was absorbed and neutralized in the sea. The dioxins were broken down in an interaction with ultraviolet light from the sun and algae. This split the compound into simpler, harmless compounds. The process took only a few days.

The *Vulcanus* burned its poisonous cargo in the Gulf of Mexico, off the Texas coast. As the incineration went on, angry protesters lined the shore. They feared that dangerous amounts of dioxin-contaminated gases would be blown inland. After all, they pointed out, the *Vulcanus* was only able to destroy 99 percent of the wastes. A full one percent remained in liquid form.

One percent may not sound like much—but remember how toxic wastes like dioxins are in even microscopic amounts. On land, the EPA rules for hazardous waste incineration are much stricter. They require that 99.99 percent of the waste be made harmless. In other words,

CANUS
SAPORE

only one part in 10,000 of any hazardous material can legally go up the stack and into the air. This may be sufficient to avoid danger, particularly since the hazardous materials will be further diluted as their smoke and gases spread through the air.

Microorganisms such as bacteria can be used to break down hazardous wastes. A typical method is the activated sludge system. As we saw in the first chapter, sludge is a soft wet mass of waste. Bacteria are added to the mass and air is bubbled through it. The bacteria form tiny clusters about one millimeter (0.04 inch) across. Each cluster contains millions of bacteria. The bacterial clusters absorb the particles of waste in the sludge around them. Using the oxygen dissolved in the water, the bacteria break the wastes down into water, carbon dioxide, and other harmless substances.

This process, and others that use microorganisms, are much cheaper than incineration. The problem is to find organisms that will "eat" some of the more toxic wastes. Or to create such organisms.

The EPA will spend about one million dollars between 1983 and 1986 on looking for organisms that can digest chlorinated organics. One of the scientists the agency has hired to do this investigation is a pioneer in the field of biotechnology: Dr. A. M. Chakrabarty. He is a microbiologist at the Medical Center of the University of Illinois in Chicago.

Dr. Chakrabarty searched through wastes from Love Canal and other toxic dump sites. He discovered several microorganisms living in—and on—the wastes. Some are able to break down 2,4,5-T.

Chakrabarty hopes to isolate the genes in the bacteria that enable it to "eat" 2,4,5-T. These genes encode the

The incineration ship, Vulcanus

"recipe" for the chemical processes the bacteria use in digesting the toxin. He will then try to transfer these genes to other microorganisms. Eventually, Dr. Chakrabarty wants to develop strains of bacteria that can break down dioxins and other chlorinated organics besides 2,4,5-T. However, very few such "tailor-made" organisms can grow or survive outside the laboratory. It may be many years before this technique is of real use in cleaning up hazardous wastes.

Sometimes, there is no choice but to break down highly toxic wastes by chemical means, a process that can be complicated and expensive. This is how 4,600 gallons of waste in an abandoned tank in Missouri had to be handled.

The waste was contaminated with fourteen pounds of TCDD. It was so toxic that it could not be incinerated or treated by biological methods. The tank wastes were divided into twenty-five batches. The first was treated with an organic solvent to dissolve out the dioxin. The dioxin solution was exposed to high-intensity ultraviolet light. This separated the chlorine from the dioxin molecules. The dioxin was broken down to hydrochloric acid—a useful chemical—and a number of organic compounds. The compounds were much less toxic than dioxin and were further broken down by biological methods. The solvent was recovered from the process, purified, and reused on the next batch, and so on. It cost $500,000 to design the equipment for this treatment. The treatment of all twenty-five batches ran into many millions of dollars.

Another waste-treatment method involves physical processes, such as adsorption. In adsorption, a substance like finely divided charcoal "soaks up" large volumes of toxic liquids or gases much like a sponge. The toxic material can later be removed from the charcoal for further treatment.

Yet, some wastes cannot be made harmless by *any* of these methods. Radioactive wastes are one example. So are heavy metals, such as lead, mercury, and cadmium. They are toxic and all their compounds are toxic. True, a part of these metals can be reclaimed from the wastes containing them. But this is a very expensive process, and there is always some metal left in the waste.

Since these wastes cannot be made harmless, they can only be stored. They end up in landfills. Like radioactive wastes, they are first sealed into the molecular structure of a stable solid. The solids generally used are cement, lime, or asphalt.

Molecules of lime or cement form a strong bond with heavy metals. But in a landfill that contains many other materials, lime and cement may be weakened by corrosive substances. Constant freezing and thawing also breaks up these "containers." Asphalt, on the other hand, is not affected by water or corrosion. But it softens and burns at high temperatures. And it can be dissolved by organic solvents.

There is no such thing as permanent storage for heavy metals. All we can do is put them as far away from our food, our water, and ourselves as possible, and keep a wary eye on them. But for practically all other hazardous wastes, there are effective treatments that will make them harmless.

That's the good news. The bad news is that most of these wastes are still being buried—and forgotten—in landfills.

CHAPTER FIVE

WHAT INDUSTRY SAYS ABOUT HAZARDOUS WASTES

"The water down there," said one Holbrook, Massachusetts, town official, "smells like Mr. Clean." The official didn't mean the water smelled *clean*. He meant it had the sharp, chemical odor of a household cleanser. The water he was talking about was the water in one of the wells the town of Holbrook uses for its public water supply.

There is no doubt that Holbrook has a hazardous waste problem. The town's chemical dump ranks eighteenth on the EPA's most-dangerous list. Among the toxins there is phenol.

That is where certainty ends for Holbrook. All the rest—how the chemicals may spread through soil and water, how they may affect people, animals, and plants, exactly how the chemicals got there in the first place, how the site might be cleaned up—is open to dispute.

The same is true of thousands of other known toxic waste dump sites. They contain hazardous wastes. That much everyone agrees on. Beyond that, disagreement begins.

Just east of Los Angeles, in Riverside County, California, lie the Stringfellow Acid Pits. Over sixteen years, from 1956 to 1972, about 34 million gallons of wastes—

The Stringfellow Acid Pits,
a toxic waste
dump site
near Los Angeles

acids, heavy metals such as zinc, mercury and lead, and pesticides—were dumped at the twenty-two-acre site.

The Caputo Pit is in Moreau, in upstate New York. It is number 140 on the EPA list. Among the pollutants there are 452 tons of PCBs and poisonous solvents. They were put in the Caputo Pit in the 1950s and 60s.

In Wayne, New Jersey, the Sheffield Brook is a favorite play area for children. Part of it has also been contaminated by radioactive wastes, which were dumped near the stream between 1948 and 1971.

In Times Beach, Missouri, dioxin levels have been measured at 100 ppb in parts of the town.

Dioxin in Times Beach . . . radiation in Wayne . . . PCBs in Moreau . . . acids and heavy metals in Riverside County . . . phenol in Holbrook . . . what do they mean—what threat do they pose—for residents of those towns? How threatening are the other 400-odd dumps on the EPA's most-dangerous list, the 15,000 or so additional hazardous waste sites in the nation, or the thousands of other dumps around the world? People do not agree.

One disagreement involves the security of the dumps themselves. In some places, there is no security at all. Just as Minimata Bay in Japan was polluted by a plastics company throwing its mercury wastes into the water, so many American rivers, lakes, ponds, streams, and bays have been contaminated by the direct dumping of solid and liquid wastes.

At other locations, dumping has been more careful. Solids and sludges have been buried in landfills. Liquids and gases have been placed in large metal drums for storage.

Such precautions, however, did not turn out to be enough to protect the environment. Storage containers have leaked, and so have the landfills. Many acres of soil, and countless bodies of water, have been polluted.

But how bad is the problem, really? Most industrial leaders say it is not nearly as bad as environmentalists make out. Industrialists agree that some sites are contaminated, but, they say, these sites are relatively small and isolated. Some can be cleaned up by reburying the wastes there. Others can be cut off from nearby water supplies by the building of "slurry walls," underground retaining walls of a semiliquid mixture of clay and water. Better, more durable storage containers can be used. The seriousness of the situation, many industrial officials believe, has been vastly overstated. Environmentalists also exaggerate the threat that hazardous wastes pose to life and health, business leaders say. The extent of that threat is another issue in dispute.

In 1980, the EPA hired a Houston, Texas, firm, the Biogenics Corporation, to study the health of families living near Love Canal. The Biogenics report was sensational. It showed that of thirty-five people tested, eleven had suffered chromosome damage. Chromosomes are structures that determine the characteristics children inherit from their parents. Damaged chromosomes mean a much-greater-than-average possibility of serious birth defects. They may also induce some cancers.

The Biogenics report scared the residents of Love Canal—and galvanized the federal government. Citing the report as evidence of the deadly danger of living in the area, government officials agreed to pay for moving more than seven hundred families into new neighborhoods. The New York state government had already moved two hundred other families out of Love Canal.

Some people, though, had their doubts about the Biogenics report. These people included independent scientists, as well as industrialists. The 1980 study, they maintained, was so poorly designed that it should never have been carried out. For example, that report included as "birth defects" conditions like jaundice, pigeon toes, and cerebral palsy—conditions which are not generally considered birth defects.

In 1981, a new study was begun. This one was conducted by the United States Centers for Disease Control, and by Brookhaven National Laboratory, in Upton, New York, and the Oak Ridge National Laboratory, in Oak Ridge, Tennessee.

The second study contradicted the first. Blood samples from fifty-five former residents of Love Canal were compared to blood samples from a "control group" of forty people from another part of Niagara Falls. People in the two groups were matched according to age, sex, and other factors, in order to make the study more reliable. The finding: Men and women of Love Canal showed no unusual chromosome abnormalities at all. They were every bit as healthy as people who had never lived near the contaminated site.

Industrial spokespeople hailed the new report. They regarded it as proof of their position—that the hazardous waste danger in this country is not nearly so bad as some people would have us think. Actually, these people contend, industrial wastes pose less of a public health threat than does cigarette smoking. Figures from the American Lung Association show that as many as 325,000 Americans die each year from the effects of smoking. Far fewer are known to have suffered physically from pollution caused by hazardous wastes.

Furthermore, people in industry point out, many so-called contaminants may not be so dangerous after all. In 1983, the Dow Chemical Company, a major producer of chemicals contaminated with dioxin, announced it was undertaking a three-million-dollar study of the material. Dow officials expressed confidence that their work would prove dioxin is much less health- and life-threatening than people assume it is. The company "will seek, through sound science, to reassure those with concerns about dioxin," said Dow president Paul F. Oreffice.

The fact that the public is so alarmed about dioxin and other hazardous wastes, industry leaders continue, also illustrates something about the role that the American news media play in spreading information.

"Those people who depend on the media for information and guidance [about hazardous wastes] have been frightened, some to the point of hysteria." So wrote *Science* editor Philip Abelson. No one in American industry is likely to argue with him.

Journalism, like any other business, is a competitive field. Newspapers, magazines, radio and television stations need readers, listeners, and viewers. How to get them? Present startling, intriguing stories—and present them in poignant human terms.

. . . The camera zooms in on a mother and her infant at Love Canal . . . Has the mother suffered chromosome damage? Has she given birth to a child who may become seriously ill at any time? Naturally, she is worried. The reporter talks with her, draws her out, and presents her fears, skillfully and professionally, to viewers. They feel her fears, and become fearful for themselves and their own families. Alarm spreads . . .

Often, industry leaders protest, that alarm is not justified by the facts. According to the 1981 CDC study, the Love Canal mother has not suffered chromosome damage. Her child is as healthy as any other child.

Industry leaders are concerned about the "bad press" they are getting. One way they are trying to combat it is to issue guidelines for reporters. The Chemical Manufacturing Association, a trade group, has drawn up a series of twenty-two suggestions entitled, "Hazardous Materials Emergency: Your Need to Know." The suggestions emphasize reporting of hazardous materials spilled in rail or truck accidents, but many would apply to reporting at dangerous dump sites as well. Some have to do with safety; others emphasize the reporter's responsibility to find out all the facts *before* writing a story.

Besides criticizing reporters, businessmen and -women sometimes find fault with politicians whom they believe are too eager to leap on the "environmental bandwagon." Politicians sometimes contribute to public confusion by condemning a particular waste substance, or by calling for a certain cleanup procedure, without thoroughly understanding the problems involved. As the *Wall Street Journal*, the nation's leading business newspaper, commented in an editorial, "Congressmen find the issue [of poisons in the environment] a terrific way to get a few headlines." Many people—in business and out of it, feel that politicians, as well as reporters, ought to think beyond headlines to the science of hazardous wastes.

Such industry sentiments may seem to be self-serving, but they also represent a big step forward. Until very recent years, few business leaders considered hazardous wastes a problem. Few thought about them at all.

It was so much easier just to get rid of wastes in the easiest and cheapest way possible. Pour them into the bay. Put them in low-cost containers, stick them in a vacant lot, and forget about them. Bury them in a poorly designed landfill, and then sell the site to a developer for building or gain a tax deduction by giving the land away. (The latter is what happened at Love Canal. After burying its wastes there for ten years, the Hooker Chemical Company sold the dump to the Niagara Falls School Board in 1953. An entire neighborhood grew up around the spot.)

Some of this dumping, although careless, was legal at the time it was done. For instance, the PCBs left at the

This neighborhood, the Love Canal area of Niagara Falls, made headlines when it was revealed that its houses were built on what had once been a dump for toxic waste.

Caputo Pit in Moreau, New York, were placed there in accordance with the law. Overall, though, the law was not followed. The federal government estimates that at least 90 percent of hazardous waste dumping in this country has been illegal.

The industries that produced the wastes have done some of the illegal dumping. But much of it is the responsibility of commercial waste hauling firms. Industries hired such firms to get rid of their waste products. By and large, the record of these haulers is dismal. It is a major reason scientists find it so difficult to track down just which companies are responsible for which dump sites.

Many commercial waste haulers store wastes in leaky containers. Others mislabel the contents, making highly toxic wastes appear to be less dangerous than they really are. Still others, when they find their storage tanks filled to overflowing, simply "pull the plug." Deadly chemicals ooze or gush out into soil and water.

Some commercial waste firms are called "midnight haulers." They load up trucks with containers of toxic poisons and head, by night, for remote or sparsely populated areas. There, they dump their loads in rivers, fields, or by the sides of roads.

Some wastes are sent out of the country. A Houston company transported barrels of PCBs to Mexico. Mexicans, not knowing what the barrels contained, emptied them and used them to store drinking water. Those who used the water became ill.

Commercial waste hauling is a big business—and a profitable one. Industries are willing to pay handsomely to get their wastes out of the way. The commercial dumpers, disposing of hazardous substances generally in the cheapest way possible, keep their expenses to a minimum. In addition, some make a second profit on the dangerous materials they are paid to haul away. For example, some have been known to resell contaminated oil as a fuel.

Today, lawyers from the United States Department of Justice's environmental crimes unit are busy looking into the commercial waste-hauling business. They are investigating the illegal dumping of past years. In the process, they have found that some of the firms have close ties with organized crime. The industry executives who hired those firms to take care of their wastes deny that they knew of the crime link. Most also say that they did not realize wastes were being disposed of illegally.

In 1979, the Justice Department began to crack down on the illegal waste haulers. By April, 1983, department lawyers had brought criminal charges before the courts in fourteen states. The first case to come to trial, in Concord, New Hampshire, ended in the conviction of a Massachusetts firm.

The federal government has taken other actions to try to protect Americans from the wastes they produce. In 1976, Congress passed the Resource Conservation and Recovery Act. RCRA (pronounced "recra") attempted to regulate the transportation, storage, and disposal of about forty million or so tons of toxic wastes each year. The law was amended in 1980, but it expired two years later. By mid-1983, Congress was working on a new RCRA. This law, says one of its congressional sponsors, needs to be even tougher than the old one.

In 1980, Congress passed another important hazardous waste bill. This one created the "Superfund"—$1.6 billion in federal money to be used to clean up abandoned waste dump sites around the country. The dumps on the EPA's most-dangerous list are slated to receive Superfund monies.

Here's how the Superfund is supposed to work: Money is taken from the fund, and the clean-up begins. Meantime, the company or companies found to be responsible for the wastes work out an agreement with the government to pay back the Superfund money. That way, the clean-up takes place—but the polluters pay the bill.

Environmentalists back the idea of the Superfund, and they are enthusiastic about a new, tougher RCRA. Industrialists are not so happy about the prospect. They wish the government would let them alone—let them take care of their own wastes in their own way. They claim that when the government tries to legislate environmental matters it makes it difficult for business to operate for the good of the country.

Environmental laws are expensive and time-wasting, industry leaders say. Their companies not only have to meet federal, state and local standards, they must also demonstrate that they have met them. That demonstration requires massive amounts of paperwork—letters, reports, forms, applications, evaluations, appraisals, and reappraisals. Many companies have had to hire new workers just to cope with the paperwork. That is expensive.

There are other expenses involved in meeting government standards. Better storage containers, deeper landfills, improved treatment of toxins; all this can cost millions. It's a burden for large companies. For small ones, it can amount to the difference between a profit and a loss that puts the firm out of business. Putting a small firm out of business means throwing tens, or even hundreds, of men and women out of work. Environmental laws can damage the nation's economy, business leaders argue.

That damage to the economy must be balanced against possible dangers of hazardous wastes, industrial spokespeople say. It should also be balanced against the benefits of allowing industry to operate with less government interference. Such a balance is called a "cost-benefit ratio." Industry leaders believe the risk of environmental damage from hazardous wastes is outweighed by the benefit of allowing industry to expand, and to offer more jobs and more products to the American public.

Cost-benefit analysis, they contend, dictates that we accept the hazards of living with a certain amount of industrial pollution.

Industry also objects to the sweeping nature of many environmental laws. Congress has set impossible clean-up deadlines, business leaders say. It is asking industry to do too much, too fast.

Many people think this argument makes sense. In 1980, a congressional subcommittee drew up a report on the effects of some of the environmental legislation of the previous ten years. "The history . . . shows that the 'asking for everything at once' approach has produced enough confusion and inefficiency to have all too often slowed, rather than accelerated, the achievement of our goals," it stated.

Despite that report, American industry knows it is under congressional pressure to "clean up its act." It is under pressure from individuals, too. Today, industry faces many lawsuits brought by men and women who believe they have suffered, physically or emotionally, from hazardous wastes near their homes, schools, and places of business. Already, such lawsuits have cost business millions of dollars.

So business has a financial reason for seeking solutions to the hazardous waste problem. And seeking solutions, business leaders say, is exactly what they are doing. The Mobil Oil Corporation, for example, has started to treat acidic waste oils at its Joliet, Illinois, plant with lime before sending them to a landfill. Previously, such wastes were buried untreated. Not far away is a United States Steel facility. Until recently, its wastes were being dumped on the banks of the Des Plaines River. Now they are being hauled to a dumpsite in Indiana.

At a nearby Dow Chemical plastics plant wastes are being incinerated rather than being buried or allowed to flow into the river. Some people at Dow are enthusiastic about incineration. "Land disposal is nothing but storage of the waste," according to the company's manager of environmental regulations, James Karl. "As long as it's in existence, it can cause trouble."

Other businesses are experimenting with the recycling and reclaiming of certain waste products. To such methods, some companies are adding matchmaking. "Matchmaking" is waste exchange. One company produces a hazardous by-product for which it has no use. But for another company, that "waste" may be a vital raw material.

How do the companies get together? Through a matchmaker—one of the three major nonprofit waste clearing houses in the United States. The three are located in Syracuse, New York; Tallahassee, Florida; and St. Louis, Missouri. Four times a year, each publishes a list of companies with "materials wanted," or "materials offered." The two lists are compared, and buyers are matched with sellers. The matchmakers are "inventing a new market," asserts Walker Banning, manager of one of the regional centers. "We're creating something whcrc there was nothing before."

American industry is acting on the hazardous waste problem. How effectively? Former EPA director Anne McGill Gorsuch Burford was confident about industry's readiness and ability to deal with its own wastes. "Private industry has increasingly demonstrated the willingness to accept the responsibility for past actions and to pay the bill for further remedial action," she said.

Many in industry sound equally optimistic. In the past, maintains Richard Symuleski, environmental coordinator for Standard Oil of Indiana, "Solid waste was a stepchild and we had no technology. Technology is emerging now." But Symuleski adds a note of warning. "We're still learning. It's going to take a long time . . ."

Environmentalists agree. If it is left up to industry, they believe, it will take a long time indeed to solve America's hazardous waste problem. As evidence, they could point to the words of H. Barclay Morley, chairman and chief executive officer of the Stauffer Chemical Company. Speaking to the Society of the Chemical Industry, Morley said, "As an industry we have maintained that . . . we will work to eliminate any health hazard or problem that is real and which has been explicitly and properly defined."

Those words, "real" and "explicitly and properly defined," may be the key to the future of American industry's attitude toward its wastes.

CHAPTER SIX

FACTS, FEARS, AND CONFUSION: THE PUBLIC'S VIEW

It's no fun living next to one of the nation's hazardous waste dumps. Air around the site may smell foul because of noxious gases rising from underground. The earth may appear slimy or iridescent. It may be covered with a thick, greasy scum. Small streams or pools of water may look or smell peculiar, too.

Worse than the smell or the appearance is knowing that the dump may be life-threatening. Is it safe to stay in my home? people ask. To open the windows in hot weather? To have a cookout? To let the kids wander around the neighborhood?

Worst of all is asking such questions—and having them go unanswered week after week and month after month. Yet that is just what is happening to a growing number of American men, women and children. They are living with toxic chemicals and they have questions. Few of them have heard any satisfactory answers.

That does not mean that they don't get any answers at all. It means that the answers they do get don't seem to help them much.

Take industry's answer, as expressed by H. Barclay

Morley to the Society of the Chemical Industry: "We will work to eliminate any health hazard or problem that is real and which has been explicitly and properly defined." What does that mean?

Finding the explicit and proper definition of a problem is never easy. It is especially difficult when the problem is as complex as that of hazardous waste.

How do you "explicitly and properly define" the way toxic substances move through soil and water? Such movement is largely unpredictable. It depends upon the nature of the substances, the slope of the land, the type of soil, the level of the watertable, and numerous other factors.

Of course, such factors can be scientifically analyzed. An industry that is considering using a particular spot for a toxic waste dump may call in experts—such as geologists and engineers—to examine the site. After carrying out a thorough study, they will recommend for or against the location. If the recommendation is positive, they will use their scientific knowledge to suggest the best design for the site. Suggestions might include the building of retaining walls or the recontouring of land in order to make the dump more secure.

Unfortunately, though, even the most careful analysis cannot take all possible factors into account. Once the dump is in operation, it might undergo an exceptionally fierce storm or unprecedented flooding. Erosion could occur. Waste materials other than those the experts considered might be left there. Any of these factors could make the site more dangerous than predicted.

Scientists know a good deal about the movement of substances through land and water, but they do not have all the answers. "It's like playing Russian roulette," says a Moreau, New York, man who lives 2,800 feet from the Caputo Pit. "You don't know where [the toxic waste] will go or what it'll do next."

CANAL
WE'VE GOT BETTER
THINGS TO DO THAN
SIT AROUND AND BE
CONTAMINATED!!

Nor do scientists know all there is to know about the effects of various toxins on life and health. This is another area where explicit and proper definition is extremely difficult. In some cases, scientists cannot even be sure what their own studies indicate.

So far, there have been two, quite contradictory, reports about possible chromosome damage to people who used to live near Love Canal. Which one should they—and we—believe?

According to several scientists, the first report contained serious design flaws. The second may be flawed, as well. One of its authors, Matthew Zack, pointed out that it, like the first, looked at only a small number of people. That may have distorted the results. Another problem is that the second group of scientists did not complete their work until as many as three and a half years after the residents had left the polluted neighborhood. Just because their study did not show chromosome damage is not proof that the damage never took place. The body can repair some kinds of chromosome damage. The people of Love Canal may have suffered damage, most of which could have cleared up naturally after they left their homes. Or, damage may still exist, but it may be so subtle that it was not picked up in the CDC tests.

Clark Heath, another of the authors of the second report, also expressed reservations about it. It is impossible to link chromosome abnormalities scientifically with birth defects, cancer, or other diseases. "The data don't even exist yet," he complained.

A young resident of Love Canal joins the public protest against the dangers uncovered on his street.

Where does that leave the people of Love Canal? With two sets of answers—contradictory answers—both of which may be wrong!

Another problem with explicitly and properly defining the health effects of toxic wastes is that, so far, few studies have been conducted over long enough periods of time. Exposure to dioxin at Love Canal in 1978 may result in cancer—but not until 2008 or 2018. It may mean a defect that doesn't show up until a baby has become a teenager.

Thirty years ago, numbers of United States military personnel were exposed to radiation during nuclear weapons testing. Government authorities assured them that they faced no health hazard. At first, that seemed to be true. Those exposed remained apparently healthy for years. But now we know that the cancer rate among those ex-service people has been greater than the cancer rate for the general population. They were poisoned by a hazardous waste—radioactive fallout—but that poisoning took years to reveal itself.

Still another factor makes it difficult to define the human health effects of toxic wastes. That factor is synergism.

"Synergism" comes from two Greek words, *syn,* which means "with," and *ergon,* "work." It refers to the combined effect of two or more substances. In a synergistic reaction, each substance will have a greater-than-normal effect because of the presence of the other.

There are many examples of synergism between two hazardous substances. Both cigarette smoking and continued exposure to asbestos dust are known to cause cancer. Asbestos is a mineral widely used in insulating and fireproofing. An asbestos worker who smokes is ten times more likely to get cancer than one who does not smoke.

There is evidence of synergism between TCDD and other hazardous chemicals. The experiments were done in 1982 at the University of Wisconsin McArdle Laboratory for Cancer Research. Mice were exposed to certain chemicals that irritate the skin and sometimes cause skin cancers. Some of these mice were exposed to TCDD as well. Many more of the mice in the second group got skin cancers. The experimenters point out that the reaction produced by TCDD in the skin cells of mice is very much like that produced in human skin cells. "There is no question that dioxin plays a role in tumor formation in animals, and almost certainly has the same action in human cells," says Dr. Howard Eisen of the National Institutes of Health.

But which hazardous waste combinations are dangerous, and in what amounts? No one is sure. One reason is that scientific tests generally concentrate on a single toxic agent at a time. Each substance is studied in isolation. Until scientists have investigated the synergistic effects of hazardous wastes in detail, it will be almost impossible to tell how dangerous our nation's dumps really are.

Even then, there will be the difficulty of relating laboratory results to real life. Sure, experiments have demonstrated that dioxin, for instance, is highly toxic. But, "There is a difference between toxicity and hazard," proclaims Etcyl Blair, a scientist at Dow Chemical Company. "Just because it is a risk doesn't mean you are in danger."

The president of the Chemical Industry Institute of Toxicology, Robert A. Neal, agrees that it is hard to decide what is risky and what is not. "Toxicology is not an exact science so there is latitude for interpretation of what the risks are," he commented in 1980.

Sometimes, that latitude has been deliberately increased. Some people suspect that is what is happening with the $3 million dioxin study announced by Dow in the spring of 1983.

Dow has produced dioxin for years. It was a contaminant in the Agent Orange the United States government bought from Dow for use in Vietnam between 1961 and 1971.

During at least part of that time, Dow officials knew that dioxin was a probable cause of birth defects. The officials admit that. They also charge that the United States government was aware of the dioxin hazard. As early as 1969, company spokespeople claim, a study by the National Cancer Institute showed that mice who have been exposed to dioxin have an increased risk of producing defective offspring.

Wait a minute. Dow officials knew as far back as 1969 that dioxin was a health hazard? Then what about the statement of the company's chairman of the board, Robert Lundeen, on April 20, 1983? Speaking on the CBS-TV Morning News, Lundeen said, "We stand on thirty years of producing and using the material [Agent Orange] and either in our plants or in the people used [sic], there is no indication that there were any adverse health effects."

Dow spokespeople claim they can say this because other companies—for example Monsanto—also made Agent Orange. And, Dow says that its product contained less than one part per million of dioxin, while the Agent Orange made by other companies had ten to fifty parts per million. As we shall see at the end of this book, scientists now generally agree that one part per billion is a maximum permissible level for dioxin in soil. That is an amount a thousand times smaller than one part per million.

There are other questions. Why is Dow willing to spend three million dollars on a dioxin study? Company officials say they plan to use the study to "reassure those with concerns about dioxin."

How reliable will that study be? How much faith will the people of Love Canal, or Times Beach, or any other dioxin-contaminated site be able to put in it? By its own admission, Dow has known for nearly fifteen years that dioxin is related to birth defects. Now the company begins a study that is designed, before it even begins, to be reassuring.

Dow is not the only company that seems willing to bend the facts in order to reassure the public. Monsanto, Inc., another giant in the chemical industry, makes a product called Aroclor 1254. Aroclor contains PCBs.

During the 1970s, Monsanto hired a chemical testing laboratory, Industrial Bio-Test, Inc., of Northbrook, Illinois, to run toxicity studies on its product. According to Bio-Test's report, Aroclor is "slightly tumorigenic." That is, people exposed to it have a small risk of developing tumors. Monsanto officials didn't like the sound of that. In a letter, they ordered Bio-Test scientists to reword their finding. The scientists complied. Aroclor "does not appear to be carcinogenic" (cancer-causing), the final report read.

This kind of twisting of fact angers many Americans. Nor do they believe that industry twisting is limited to instances involving the definitions of health hazards. They think it comes up again and again in matters of hazardous waste. They see it, for example, in industry's use of landfill dumping. Other, more sophisticated, technologies are available, and some companies are experimenting with them at one or two plants. An example is the method of breaking down dioxin mentioned in chapter four. But so far, industry in general remains committed to landfill.

Why? Many people are convinced that the answer boils down to money. Industry favors landfill, they say, because it is so much less expensive than incineration or other, more advanced, technologies.

That is not true, industry leaders protest. They do not prefer landfill because they want to save money. Their preference is based on the evidence of scientific cost-benefit analyses. Such analyses indicate that the risk posed by leaking landfills is not great enough to justify spending the huge amounts that would be needed to destroy or recycle all their wastes.

Environmentalists respond by asking how reliable cost-benefit studies really are. No study is more dependable than the data upon which it was based. For example, Dow Chemical, in announcing its 1983 dioxin study, also announced what the study's outcome was going to be. Clearly, "knowing" in advance how the study will turn out is going to affect the way Dow scientists collect, study, and interpret their data. Similarly, industry scientists who already have a strong commitment to landfill dumping are almost sure to come up with "proof" that this method is perfectly safe. In fact, they have already come up with it. In the words of the Chemical Manufacturers Association, "landfilling of most hazardous waste can be done in an environmentally sound manner."

Few people outside industry would concur. According to an article in *The New York Times,* "all experts agree," that "even the most advanced current landfills will eventually leak."

Dr. Samuel Epstein, environmental scientist and author of a book on hazardous wastes, is one who thinks this way. "The use of secure landfills is not the way to go," he says. "They're not really secure. In fact, they're impractical and unsafe . . ."

This opinion seems to be borne out by the facts. A Texas scientist studied newly developed landfill "liners," which are designed to prevent leakage. In the study, eleven out of twelve liners leaked after just six months. In a test of four New Jersey landfills, each equipped with two separate liners, leakage began within a year. "I think the whole idea of secure landfill is really a figment of optimistic imaginations," said the scientist who conducted the New Jersey study.

Industry leaders claim that the leakage problem can be controlled. Landfills will be monitored for up to fifty years before being permanently capped. But as we have seen, many toxic wastes will remain dangerous for many times fifty years. Toxic metals will be poisonous *forever.*

The arguments rage back and forth. And the public waits.

The people who used to call Love Canal "home" wait to find out whether they, and their children, have been harmed by dioxin. They wait in fear. "I'm horrified," says the mother of two children, one of whom has a birth defect.

The people who used to live in Times Beach wait, too. In June, 1983, federal government officials gave the final approval to plans to buy the town from its residents. But that doesn't solve the residents' problem. Many thought the government was giving them too little money for their homes. Few wanted to face the upheaval of moving. Even when they are settled in new homes, they know the dioxin threat will not be over for them. It may not be over for thirty or forty years—however long it may take for a possible illness to develop.

A few Times Beach people decided to do their waiting right in their own dioxin-contaminated town. It had become a ghost town—no running water, no police protection, no mail delivery. People had to carry their garbage a mile down the road to a pickup point. A few children played in the lonely streets. "They do miss their friends," said one mother.

People are waiting in Globe, Arizona, too, in an asbestos-contaminated trailer park. They're waiting in Moreau, in Holbrook, in PCB-contaminated Waukegan, Illinois, and near the Stringfellow Acid Pits. They're waiting near every one of the EPA's 418 most-dangerous toxic waste sites across the country, and near thousands of other sites as well.

We are all waiting. No one knows where hazardous wastes will appear next.

Suppose it turns out to be your home town. What will you do? To whom will you turn for help?

For many people, the answer to that last question has been, "The government." And the government has been involved in many toxic waste emergencies. It has sponsored studies, provided housing, helped people find and pay for new homes. Yet critics say it can—and should—do more.

Government involvement in the nation's hazardous waste problems has been too piecemeal, its critics say. It has concentrated on this dumpsite, on that town, on this neighborhood or street. It has not acted forcefully to compel polluters to clean up their own wastes. It has done little to ensure that America's future toxic waste problems will not be even worse than today's.

In 1980, a committee of the House of Representatives ordered the Congressional Office of Technology Assessment to begin a study of federal hazardous waste guidelines. That report was released in March, 1983. Its conclusion: Government rules do not assure "protection for human health" from "massive annual accumulation of hazardous waste."

The report was critical of the EPA. In the first place, it said, the agency vastly underestimates the amount of waste produced each year. That may amount to as much as 275 million metric tons a year—more than one ton for every citizen of the United States. The EPA estimate is about 40 million tons a year.

The report also faulted the EPA for continuing to endorse landfill dumping. It is "highly probable" that "hazardous constituents" will eventually leak from the nation's landfills, the report concluded.

The Office of Technology Assessment report also referred to EPA's pattern of "delay, false starts, [and] frequent policy reversals." It could, although it did not, also have mentioned that the agency has made decisions based on politics, rather than on environmental and health needs.

During the winter of 1982, a major scandal developed at EPA. Part of the scandal centered around the head of the agency, Anne McGill Burford.

Burford became EPA administrator in 1980. She was appointed to the position by President Ronald Reagan.

At once, critics charged, Burford began a campaign to get rid of environmental activists at the agency. Some were fired. Others were given little work to do, or were treated so unpleasantly that they eventually resigned. Their places were left vacant, or filled with men and women who had worked for industry in the past.

One of these people was named Rita Lavelle. Before coming to EPA, she was director of public relations for a branch of Aerojet General Company in California. Aerojet General manufactures jet and rocket engines. It is a major user of toxic waste dumps. At EPA, Lavelle was put in charge of administering the $1.6 billion Superfund for toxic waste cleanup.

Lavelle's handling of the Superfund was another part of the EPA scandal. Six million dollars of that money was supposed to be used to start cleaning up the Stringfellow Acid Pits. It was Lavelle's job to get the cleanup going, and then to work out an agreement with the Stringfellow polluters to pay the government back.

One of those polluters, however, was Aerojet General, Lavelle's former employer. Apparently, Lavelle and officials from Aerojet—her former bosses—worked out a "sweetheart deal" on the Pits. By that deal, Aerojet would get out of paying most of its share of the cleanup costs. The part the company played in polluting the Stringfellow Pits would be kept secret. When members of Congress got hints of what was going on, they launched an investigation. Lavelle lost her job. (Some months later, in December, 1983, Rita Lavelle was convicted of

lying to the congressional committee that investigated her activities as Superfund director.)

But there was even more to the EPA scandal. Members of Congress learned that Lavelle, Burford, and others had conspired to delay the start of cleanup operations at Stringfellow in 1982. The reason? They wanted to help a Republican from California get elected to the United States Senate that year. The election contest was between a Democrat, Jerry Brown, and a Republican, Pete Wilson. Brown was governor of the state at the time of the election. If Stringfellow remained dangerously polluted, the people at EPA believed Californians would blame Brown, not the federal government. That, they hoped, would hurt his chances in the election. Their hope was fulfilled. Brown lost the election.

At the same time the EPA was dragging its heels in California, it was speeding up the cleanup of two waste sites in New Jersey. Their aim there was to help the Republican candidate for governor, Thomas H. Kean. Kean won the election.

The goings-on at the EPA shocked members of Congress. A few weeks after Lavelle left the agency, Anne Burford also was forced to resign. In her place, President Reagan appointed William Ruckelshaus. Ruckelshaus had been head of the agency earlier, from its formation in 1970 until 1973.

The EPA scandal shocked the nation, as well. Most Americans are aware that toxic wastes are a deadly hazard. They are not a matter for political game-playing. They must be treated seriously, not ignored for the sake

Rita Lavelle is testifying before a Senate committee investigating her alleged misconduct in administering the EPA's Superfund for toxic waste cleanup.

of politics or in order to save money for business. They must not be brushed under the rug by flawed or fraudulent studies.

We, the public, want a real start to solving our hazardous waste problem. We want a cleanup of the old abandoned sites, the "ticking time bombs," that haunt us all. We want to be sure no new hazardous dumps are created.

And we want answers to our questions: How dangerous are the dumps? How toxic are materials like dioxin? Is my family safe?

Who will provide the answers?

No one—not if we insist that those answers be explicit and definitive. Such answers do not exist. Industry cannot give them to us, although it claims to be trying.

Nor can science. No scientific study will ever be able to show exactly how dangerous each toxic waste is. No scientist will ever demonstrate that a precise number of people will suffer in a precise way because they once lived a precise number of feet away from a hazardous waste site.

If we wait for such "explicit and proper" answers before seeking solutions to our hazardous waste problem—as industry might have us do—we will wait forever.

William Ruckelshaus, shown here with EPA employees, was appointed to head that agency for a second time, in the wake of the scandal that resulted in the resignation of Anne Burford in 1983.

CHAPTER SEVEN

WHERE DO WE GO FROM HERE?

Who is responsible for the hazardous waste mess?

Who is going to clean it up? Who is going to pay for cleaning it up?

How can we prevent the problem from getting so bad again?

These are simple questions, easy to ask. But very hard to answer.

Why? Let's trace the story of one hazardous waste disaster from its beginnings.

It is the fall of 1971. Judy Piatt owns and runs a large stable. She loves horses. But now dozens of her horses are dead or dying. First, they lose weight and hair. Then they go into convulsions. Then they die.

Why were they dying? There had been no change in the horses' diet. Their symptoms were not like any disease of horses that Mrs. Piatt had ever heard of. Nothing had changed at the stable.

That was not exactly true, she recalled. Not long before the strange sickness appeared, the dirt floors of the stable were sprayed with waste oil to keep the dust down.

Waste oil is oil used in lubrication or manufacturing that has become contaminated with impurities. Now some industries clean up such waste oil and reuse it. In the early 70s, the oil was simply dumped.

At that time, there were no federal or state laws classifying waste oil as a hazardous substance. Spraying for dust control was not uncommon. Today the Federal EPA and many states do classify waste oil as a hazardous substance.

Oddly enough, the man who sprayed Judy Piatt's stable raised horses, too. He had a stable of prize-winning show horses. But his main business was oil salvage. He collected used oil for resale to refineries and for fuel. Occasionally, he was hired to spray oil on streets and at other sites for dust control. As a sideline, his trucks hauled chemicals for some large companies.

Mrs. Piatt wrote to state officials about her problem. They sent out a young veterinarian, Patrick E. Phillips. He and another veterinarian, Arthur A. Case, confirmed that there was something in the oil that was killing the horses. Whatever was mixed with the oil was powerful. Dr. Case said it "would knock your hat off, make your nostrils burn and your ears ring." The vets sent samples of the oil-soaked soil to the Centers for Disease Control in Atlanta.

Other stables in the area were losing horses too. All had had their floors sprayed by the oil salvage dealer's trucks.

Mrs. Piatt began following the trucks in her car, often in disguise. That went on for fifteen months. During that time, she saw the trucks spray oil at sixteen different places. They also dumped chemical wastes.

Judy Piatt sent a list of spray and dump sites to federal and state officials. There was no response.

It was now 1973. The CDC had had the soil samples for two years and there was no report from them. But scientists there were working on the problem. They had no clue as to what the toxic substance in the oil might be. There can be a lot of poisonous things in used oil—heavy metals and all kinds of dissolved organic compounds.

The search went on. In 1974, some long colorless needle-shaped crystals were discovered in soil samples from the Piatt farm. They were crystals of trichlorophenol (TCP).

That was a breakthrough. TCP is used in the manufacture of 2,4,5-T, and of hexachlorophene, a germicide. More importantly, TCP, 2,4,5-T, and hexachlorophene are often contaminated by the dioxin TCDD.

By August, 1974, CDC scientists knew that there was dioxin in the soil from the Piatt stable. There was a *lot* of dioxin—more than enough to kill the horses. The samples had 31,000 to 33,000 ppb. That is more than 30,000 times the maximum "safe" amount of TCDD as determined by the CDC.

(It's important to note that this "huge" amount of TCDD is still very small by everyday standards. It is a little more than three parts per million—or about one drop in four gallons of water. Practically all experts agree that this is a highly dangerous level for animals and people alike.)

Now the pieces of the puzzle were starting to come together. The used-oil dealer also hauled waste chemicals. Somehow, a part of that waste—loaded with dioxin—had gotten mixed in with the oil he sprayed.

But where had the dioxin-contaminated waste come from? The CDC told Dr. Phillips what it had found, and sent along two of its own doctors to help uncover the trail.

One clue they had was that TCP is used to make 2,4,5-T, an ingredient in Agent Orange, which was made for the Department of Defense. A search of Defense Department records showed that a chemical company in a town called Verona had once manufactured Agent Orange. Verona was not far from the Piatt stables and some of the other poisoned sites. But—*that company had never hired the used-oil dealer to haul its wastes.*

The doctors looked further. The company that made Agent Orange had sold its plant to another company. That company, in turn, had rented a part of the plant to yet another company, North Eastern Pharmaceutical and Chemical Company (Nepacco).

Nepacco, the investigators found, had manufactured hexachlorophene at the plant. As we've seen, hexachlorophene is a source of dioxin contamination.

Nepacco had abandoned the plant years ago. But the investigators found an old waste storage tank left on the plant site by Nepacco. It held 4,600 gallons of sludge contaminated with about 330,000 ppb of TCDD—ten times the concentration in the Piatt stables.

Thus, wastes produced at the Nepacco plant were likely to be heavily contaminated with dioxin. *And Nepacco had hired the used-oil dealer to haul their wastes as early as 1971.*

Where else had the poison been scattered? Judy Piatt had traced sixteen sites where oil had been sprayed and chemicals dumped. How many others might there be in the state? How dangerous was it to public health?

Phillips reported his fears to his boss at the state Division of Health, Dr. H. Denny Donnell, Jr. Dr. Donnell replied, "Since most of the information contained in the report was confidential, and the resulting hypothesis still unproven"—a reference to the concern over effects on public health from known and unknown dioxin sites—"it is suggested that extreme care be taken in any release of information for public consumption to the press."

Or, in plain English: "We can't be sure what the dangers are or how great they are. So tell the public as little as possible."

But that was not possible. The danger was spreading,

and there was no way to keep it secret—even if Phillips had wanted to. It was spread by an innocent bystander—a contractor who wanted some dirt to use as landfill for housing. There was dirt available at a low price from a horse arena that had closed down after a number of the horses had died. So Vernon Stout, the contractor, dug up some 850 cubic yards of it. He used most of it around a trailer park that he owned. He sold a few truckloads to some friends, Valerie and Harold Minker, for landscaping around their house.

Vernon Stout knew nothing about the TCDD in the soil he bought. Like the rest of the public, he was in the dark.

But not for long. In August, 1974, Phillips told Stout and the Minkers about the dioxin contamination. There were about 300 ppb in the soil Stout had used around the trailer park. There were about 400 ppb in the soil in the Minkers' front yard.

The Minkers decided to stay where they were. Phillips blames himself for not telling people who lived downhill from the Minkers about the dioxin contamination. He thought the soil would not erode and the dioxin would not be washed downhill. But later that happened, and the TCDD spread to other plots, and to a creek. Over the years, some of the people living nearby suffered heart, liver, and kidney damage, as well as other health problems. They believe TCDD is responsible.

On March 31, 1975, the CDC recommended to state authorities that the soil at the Minker and Stout sites be dug up and buried deep in a remote landfill. They also said that the Verona tank should be treated to destroy the dioxin it contained. However, the report mentioned that half the dioxin would break down in a year and that each year after that, half of the remaining dioxin would break down.

State officials turned down the CDC's suggestions for treating the contaminated soil and sludge. They seized on the idea that the dioxin would break down fairly rapidly. Because of that, they said, "the chance of human contact [with TCDD] appears to be very much less [each year]." They suggested more study.

The next year, 1976, Dr. Phillips came back to the Minker/Stout sites and took more samples for the CDC. The dioxin level remained the same. And the following year, the levels were still unchanged.

In fact, the scientists at CDC had made an honest mistake. Reports from the chemical industry and other sources indicated that TCDD in soil would break down fairly rapidly. This turned out not to be true.

It was not until 1977 that the state passed its first hazardous waste law. Even then, lawmakers were fearful of driving some industries out of the state by too-rapid enforcement of the law. That would mean lost taxes for the state, lost jobs for the people of the state—and lost votes at the next election. So they made sure the law would not require industries to live up to it until 1980.

And so things dragged on. In 1982, tests showed that the soil at the Piatt horse stable was just as contaminated as it had been eleven years before. The state issued its first public report on the dioxin danger. As the year ended, floods spread the contaminated soil. The public was scared. Verona, and Times Beach, and Missouri made headlines and TV news. A reluctant, scandal-ridden EPA offered to buy the houses and land at Times Beach from the residents. The original idea was to turn the area into a landfill for the dioxin.

Looking back on what happened, Dr. Phillips said, "If I had known all the heartache we would go through . . . I would have changed priorities. But what would we have done? We had no law, no money, no personnel, no incinerator. We could not force people to evacuate . . ."

Many officials and environmentalists believe that what happened in Missouri happened elsewhere

throughout the country in much the same way during the 1970s. There was no law. There was little money. Little was known about the problem of toxic wastes. And the experts agreed about even less.

So who is responsible for the hazardous waste mess? Let's consider some candidates:

Industry. Practically all hazardous wastes come from industry. By careless dumping, industry has created the huge cleanup problem we now face.

Government. Until recent years there have been few state or federal regulations on hazardous wastes. Even today the federal government, and the state governments, are moving very slowly. One result: Until recently most dumping and careless disposal of hazardous wastes has been perfectly legal.

The public. Public opinion polls show people are very concerned about hazardous wastes. They also show that people do not want any kind of hazardous waste treatment plant anywhere near them. Yet at the same time, people do not want to give up the convenience of inexpensive plastics, wood preservatives, paints, pesticides, and other materials whose manufacturing processes have toxic wastes as by-products. People send very mixed signals to their leaders. In 1980, they voted for a president clearly hostile to hazardous waste regulation.

Scientists. People know that scientists connected with various industries nearly always testify in favor of those industries. Scientists with nuclear power companies tes-

Judy Piatt initiated the investigation that revealed the presence of dioxin in Times Beach, Missouri. She angrily watches as trucks bring contaminated flood debris from there to a landfill near her new home, eighteen miles away from Times Beach.

tify that nuclear plants are safe. Scientists with chemical companies say the danger of hazardous wastes such as dioxin have been exaggerated. The public, already distrustful of industry and government, is now becoming distrustful of scientists.

It looks as if all the candidates are part of the problem. So they must all be part of the solution.

Industry must stop fighting every inch of the way to delay cleaning up hazardous wastes.

Government needs to set up some consistent standards for hazardous waste cleanup. In the summer of 1983, William Ruckelshaus, the head of the EPA, said the government needs to establish a commission to measure health risks from hazardous wastes and to determine how much should be spent to reduce those risks. He pointed out that "There is no question the air and the water are appreciably cleaner than they were in 1970—much cleaner than they would have been if there had been no [hazardous-waste] laws." Ruckelshaus said the commission should be made up of people from all walks of life who are trusted and unbiased.

The public has to decide what it will be willing to give up in the way of convenience. That is part of the cost of controlling hazardous wastes, too.

Scientists, regardless of their connections with industry or other special interest groups, should come to some agreement about the level of danger from the worst hazardous wastes.

Will any of this happen? There are some small beginnings. Perhaps most encouraging, scientists from industry and government have agreed on a standard for dioxin. On June 30, 1983, they agreed to set a standard of one part per billion as a maximum permissible level for dioxin in soil. These scientists reached agreement in spite of conflicting interests and uncertain evidence. That is what all of us must try to do. It is the first step in reducing the hazard of hazardous wastes.

FOR FURTHER READING

Brown, Michael. *Laying Waste: The Poisoning of America by Toxic Chemicals.* New York: Pantheon, 1980.

Commoner, Barry. *The Closing Circle: Technology, Nature and Man.* New York: Alfred A. Knopf, 1971.

Epstein, Samuel S., M.D., Lester Browne and Carl Pope. *Hazardous Waste in America.* San Francisco: Sierra Club Books, 1982.

Useful pamphlets from the U.S. Environmental Protection Agency, Office of Water and Waste Management, Washington, D.C. 20460, include:

"Everybody's Problem: Hazardous Waste" (SW-826)

"Hazardous Waste Facts" (SW-737)

"Waste Alert! A Citizen's Introduction to Public Participation in Waste Management" (SW-800)

More up-to-date information may be obtained from EPA regional solid waste offices around the country.

Industry's viewpoint on toxic wastes is well presented in the monthly issues of *ChemEcology*, available free of charge from: Chemical Manufacturers Association, 2501 M Street N.W., Washington, DC 20037

INDEX

ABOUT THE AUTHOR

Malcolm E. Weiss was born in Philadelphia and graduated from the University of Chicago. A noted author of science books for young people, Mr. Weiss has also worked as a science editor. Several of his books have been selected by the National Science Teachers Association as Outstanding Trade Books for Children. He and his wife, Ann E. Weiss, live with their two daughters in North Whitefield, Maine.